IMAGES OF ENGLAND

MINING IN CORNWALL

VOLUME TWO
THE COUNTY EXPLORED

Harvey's Shaft at Tresavean Mine, c.1928. The former pumping engine stack was considerably heightened to provide the necessary draught for high-pressure Lancashire boilers to drive electrical generators. The hoisting was by a large steam horizontal winding engine.

IMAGES OF ENGLAND

MINING IN CORNWALL

VOLUME TWO
THE COUNTY EXPLORED

J.H. TROUNSON AND L.J. BULLEN

The History Press

First published in 1999 by Tempus Publishing

Reprinted in 2010 by
The History Press
The Mill, Brimscombe Port,
Stroud, Gloucestershire, GL5 2QG
www.thehistorypress.co.uk

Reprinted 2012

ISBN 978 0 7524 1708 0

Typesetting and origination by
Tempus Publishing Limited
Printed and bound in England

Wheal Peevor, about a mile north-east of Redruth, was a very rich tin mine in its day. The mine was idle by 1890, around the time this picture was taken. This picturesque group of three engine houses, now roofless, can be seen from the bypass when approaching Redruth from the east.

Contents

Preface

Mining in Cornwall (Volumes One & Two) were first published in the early 1980s and republished at a later date. They have both been out of print for some years.

This revised and enlarged edition, in a different format, once again draws upon the mining photograph archive of the late J.H. Trounson, but now includes some of my own in addition.

This second volume contains a number of pictures of the Central Mining District but generally widens its focus to cover some important workings in the rest of the Duchy of Cornwall. The combined collection in total exceeds 7,000 images.

I would like to acknowledge the encouragement and support of my publishers.

The narrow but rich lode in the No. 5 Level South, at Castle-an-Dinas Mine. The lode is the vertical white quartz vein behind the miner's pick and the spots and black 'figs' in it are lumps and pockets of rich wolfram. The figure is the late J.H. Trounson.

One

St Just to Hayle

St Just, in the extreme west of Cornwall, is one of its oldest mining districts, if not the oldest, and is widely believed to be the home of underground mining in the West. As early as the sixteenth century the first attempts were made to work the tin lodes that outcropped on the cliffs, by means of adits or tunnels driven inland from just above sea level. As the workings progressed shafts were sunk to improve ventilation and for access and, as pumping machinery became available, the shafts were sunk below adit. Small mines predominated on the coast, in an almost unbroken chain running from south of Cape Cornwall to Morvah, Zennor, St Ives and Lelant. Here and there were a few larger concerns, such as St Just United Mine, which was in 1869 the first limited liability mining company in Cornwall to pay a dividend. Larger still were Botallack Mine and Levant Mine, later part of Geevor Mine. Further east lay St Ives Consolidated and the Providence Mines. Hayle, meanwhile, was the location of Cornwall's largest engineering works as well as small-scale tin mining and considerable copper mining.

A mid-nineteenth-century view of the working Cape Cornwall Mine. On the left is the pumping engine house and on the right the headgear and winding engine house. Note the long flue from the winding engine boiler house to the stack on top of the Cape.

The engines of the St Just United Mine on the cliffs above Priest's Cove, around 1863, with local fishing boats in the foreground hauled up for safety above high tide mark by means of a capstan. In the distance can be seen Land's End.

These small headgears, operated by a horse whim (a winding machine powered by a horse), were photographed near the St Just United Mine in the late nineteenth century. The trenches in the broken ground are reputed to be a Bronze Age burial site.

Bailey's Engine Shaft of St Just United is in the foreground and the engine houses of Cape Cornwall Mine can be seen on the southern side of the Cape, to the left of the picture. The headgear and pumping engine house of the latter mine are visible at the south-western tip of the Cape. The little chimney stack on the highest point of Cape Cornwall is a well-known landmark for ships. The draught from this stack, high above the boiler house, was so excessive that it was later found necessary to build another short chimney on the hillside just above the engine house. In this very old photograph, c.1863, the sea appears peculiarly flat and motionless, the reason being that the picture was taken on one of the old 'wet-plate' negatives, which required such a lengthy exposure that the waves moved an appreciable distance during the time that the camera shutter was open.

Cape Cornwall again, 1870s, but this time at low tide. The boiler flue from the beam winding engine (left of centre) can be seen climbing up over the hill to connect with the stack. The building to the right of the winding engine is the 'Count House', or office of the mine, and the small round house with a conical roof, still further to the right, is the gunpowder house or magazine, which was usually built in this form on Cornish mines.

An 1880s view of the Wheal Boys section of Wheal Owles. The town of St Just can be seen in the background.

Bellan Mine, Cot Valley, St Just, *c.*1943. In the side of the hill can be seen the tramway and chute to the mill when the plant was being installed.

The Crowns section of the famous Botallack Mine, thought to have been photographed in the 1860s. In its day it was regarded as one of the wonders of the world and even today the ruined engine houses are a remarkable sight. The lower engine house is that of the pumping engine on the Crowns Engine Shaft (this was a singularly dry mine in spite of its situation beneath the ocean). The engine house in the centre of the picture contained the winding engine that wound from the celebrated Diagonal Shaft, the top of which was further down the cliff face. This was sunk for 2,616ft at an angle of 32 ½ degrees from the horizontal so that at its greatest depth it was 1,360ft below sea level, the workings extending 2,580ft beyond the cliffs. Among the illustrious visitors to the workings were Queen Victoria in 1846 and the Prince and Princess of Wales (or, as Cornishmen would prefer to say, the Duke and Duchess of Cornwall) in 1865.

The timber staging supporting the rails on which the 'skip' ran from the winding engine down to the mouth of the Diagonal Shaft, which appears as a small black hole in the cliff at the bottom end of the staging. This scene is thought to have been photographed in the 1860s.

A memorable day in the long and colourful history of Botallack Mine was 24 July 1865, for it was then that the Prince and Princess of Wales (later King Edward VII and Queen Alexandra) descended the Diagonal Shaft and examined some of the deep workings of this famous tin and copper mine. This photograph does not show the royal visitors but the other members of the party who went down in the next skip. Note the decorative flowering bushes arranged for the occasion! Unfortunately the bottom of the negative must have been damaged and the photographer has retouched it so that the iron rails on which the skip ran do not appear in front of the vehicle. Everything was perched on or over the cliff edge, even the capstan on the right for raising and lowering heavy items in the pumping shaft.

Before reliable steel wire winding ropes became available iron chains were commonly used for hoisting in mine shafts. In April 1863 the chain used in the Diagonal Shaft at Botallack Mine broke and nine miners met their death at the bottom of the shaft. This disaster led to the painting of a famous picture entitled 'From under the Sea' by C.J. Hook, RA, that was exhibited at the Royal Academy in 1864. The original is now in the Manchester Art Gallery. To paint such a realistic picture the artist must have actually visited the scene. It depicts miners in the wheeled skip who have just arrived back safely at the surface after the lengthy journey up through the Diagonal Shaft and who are being welcomed by a young wife and her children – melodrama in the best Victorian tradition! (Photograph: Manchester Art Gallery)

Botallack Mine, c.1880. The Crowns sections can be seen in the background. Wheal Hazard shaft (centre right) had by this time had the headgear removed and the skip road carried up the cliff.

The earliest working of Botallack is believed to have been at least as long ago as 1721, but the mine finally closed down in March 1895 at a time when the price of tin had fallen to an exceptionally low level. This photograph was taken when the machinery on the Crowns section of the mine was being dismantled and forms a sad contrast to the earlier pictures taken in the 1860s.

In 1907 the Botallack mines were reopened and it was decided to lower an electric sinking pump in Wheal Cock Engine Shaft on the other side of the headland. The top part of the shaft was very close to the cliff and to make it secure it was necessary to build a massive masonry wall to a considerable height and to form a circular 'collar' for the shaft made of stones from the old engine house. The photograph shows the work in progress, the foundations of the old building visible just beyond the new collar.

The Botallack Mine power station at the time of reopening in 1907. The plant consisted of three vertical gas engines driving alternators.

A busy scene on the surface at the time of the reopening of Botallack, c.1908. From left to right are: the new mill for the crushing and concentration of the ore, the small temporary sinking headgear on the central Allen's Shaft, the roof of the power station and, in the foreground, the foundations for the new steam winding engine with the walls of the engine house being built around it.

Assembling the crushing and concentrating plant in the new mill at Botallack, c.1908.

Building the new labyrinth of arsenic flues at Botallack Mine, *c.*1908.

A group of mine officials and visitors at Botallack Mine, during the reworking of the mine (1907-1914). The gentleman on the left on the top step is Capt. William Thomas of Perranporth. He was a well-known mining personality of that time and had been the vice-principal of the Camborne School of Mines.

Levant is one of the great tin and copper mines of Cornwall, and extends for a mile beneath the ocean. This photograph was taken before Pendeen lighthouse was built on the distant headland and is thought to date from the late nineteenth century. The small whitewashed engine house, just left of centre, contains the old beam winding engine which, after the mine closed down in 1930, was the first item to be preserved by the Cornish Engines Preservation Society – now the Trevithick Society. This piece of machinery, probably the oldest steam engine in Cornwall, has now passed into the care of the National Trust.

An early photograph of Levant Mine showing the machinery that drove the man engine at this famous mine.

The old Cornish stamps for crushing the ore and the rotative beam engine which drove them, Levant, 1890s.

A late-nineteenth-century view of the Levant shafts and engine houses from the north-east. In the foreground is the inclined tunnel through which the ore crushed in the plant seen in the background was conveyed in wagons up to the stamps for fine crushing.

Major Richard (Dick) White, in the top hat, was a legendary figure in the St Just-Pendeen area who dominated the affairs of the great Levant Mine as both its purser and manager from 1871 until his death in 1909. (The purser was the secretary of a cost-book company.) Among other attainments it is said that his particular brew of punch was so potent that the smell of it a quarter of a mile away would knock any man blind drunk!

A later photograph of Major White standing in front of the doorway of the Levant Count House and surrounded by some of the members of staff.

Left of centre is the short headgear on Skip Shaft, where all the ore was hoisted at Levant, and on the right is the Cornish pumping engine and its boiler house. In the right-hand foreground can be seen the old-fashioned hand capstan for lifting heavy pump parts in the engine or pumping shaft. The two shafts were sunk so close to one another that there were only a few feet of rock between them.

An early twentieth-century photograph of the headgear on Guide Shaft at the Higher Bal section of Levant Mine and the rotative beam engine which did both the winding and pumping. It is said to have been a very economical engine to work.

A photograph of Levant Mine, taken in the first decade of the twentieth century, showing, from left to right: the house of the old beam winding engine with the headgear on Skip Shaft partly hidden by the chimney stack, the 45in-cylinder Cornish pumping engine, the capstan and, in the distance, the Cornish stamps engine. As in the case of the Botallack Mine, Levant was an exceptionally dry mine, the total amount of water to be pumped being less than 100 gallons per minute!

Men at work underground in Levant Mine.

Hand-drilling holes before charging with explosive at Levant Mine.

One of the few examples of ponies being employed underground in Cornwall for hauling wagons of ore. This photograph was taken at Levant Mine and was probably at the 278-fathom level, which extends for a mile beneath the sea.

The 'dry' or miners' change house at Levant Mine where the men washed and changed and their underground clothes were dried above hot pipes, seen here around the turn of the twentieth century.

A late photograph of Levant Mine taken in 1919, a few days before the man engine disaster that resulted in the death of thirty-one miners. The mine was partially re-equipped with new plant and struggled on for a few more years but finally succumbed to the disastrous fall in the price of tin in the worldwide slump of 1930. After a lapse of many years the old mine was absorbed by the neighbouring Geevor Tin Mines Ltd. That company later began the reopening of Levant and in order to do so sank a major flat incline shaft below the deepest workings of Geevor. Apart from the fact that the new incline commenced at the bottom of an existing mine instead of at the surface it was somewhat reminiscent of the celebrated Diagonal Shaft at Botallack Mine.

An early twentieth-century picture of miners at the Wheal Carne section of the Geevor Tin Mine.

Victory Shaft at Geevor Tin Mines Ltd, *c*.1920. The compressor house, steam winding engine house, boiler house, head gear and steel chimney are pictured. A steam lorry is delivering coal which it had hauled from Penzance.

An experimental smelting plant at Geevor Tin Mines Ltd, *c*.1920.

The electric winding engine house, ore bin and headgear of Wethered shaft at Geevor Tin Mines Ltd, c.1920.

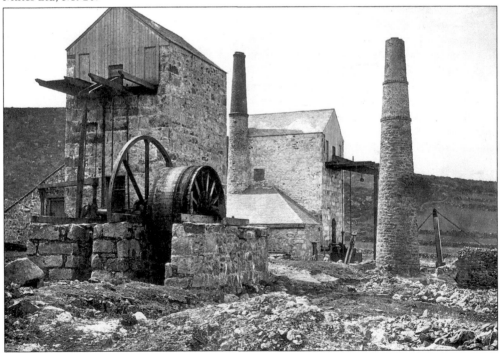

The old beam winding and pumping engines of the Carn Galver tin mine standing by the side of the coastal road between Zennor and Morvah, c.1880. The ruins of these old buildings are still standing.

Rebuilding the stack of the Wheal Mary pumping engine after it had been struck by lightning on 31 March 1886. Wheal Mary formed a part of the extensive Wheal Sisters tin mines about 2 miles south of St Ives.

A group of mine captains outside Wheal Sisters Count House, or mine office, c.1886.

The Giew tin mine was worked beneath the granite of Trink Hill, 2 miles south-south-west of St Ives. This was one of a group of old mines reopened by St Ives Consolidated Mines Ltd in 1908, but it was the only one that was successful. It continued in production until 1923 but the low price of tin then prevailing brought the enterprise to an end. The photograph, taken around 1912, shows Frank's Shaft, the principal one at Giew Mine. From left to right are the power station, the old pumping engine house (then used as an ore bin), the headgear and winding engine and the air compressor house.

A small wooden headgear on one of the lesser shafts at Giew Mine, c.1909. This shows clearly how rock and ore were raised in iron 'kibbles'.

An electric winding engine at Giew Mine, *c.*1909.

The old-fashioned dressing floors where the tin ores were concentrated, at the St Ives Consols Mines, three-quarters of a mile south-west of the centre of St Ives. This photograph was taken around 1870 and shows how labour-intensive the process was.

It is not often realized that during the nineteenth century Cornwall was one of the most industrialized regions of England. To serve the needs of the mining industry three large and several smaller engineering works were established, the largest and most important of them all being that of Harvey & Co. Ltd at Hayle, who were noted for their massive beam pumping engines. This photograph shows a large cast-iron beam (or 'bob' as they were known in Cornwall) in the works with the Hayle railway viaduct beyond, taken around 1890.

Another notable Harvey product, the pneumatic stamps for the fine crushing of tin and other ores. This set was made in 1888.

In addition to mining machinery Harvey & Co. became involved in heavy engineering in general and also in shipbuilding. This picture shows their largest vessel, the steamer *Ramleh*, of just under 4,000 tons net and built in 1891, just before leaving the fitting-out berth, with Harvey & Co.'s boiler shops in the background.

Wheal Lucy was a small tin mine in the sandhills north of Hayle. The photograph shows the chimney stack and part of an old beam engine house which had been incorporated in a smaller building for a horizontal steam engine providing power for both pumping and winding.

The valley of the River Hayle, south of St Erth, contains a shallow alluvial tin deposit which was worked during the late 1920s. The excavating was done by means of a small steam shovel which is believed to have been one of the last steam-driven shovels made in the world before the all-conquering diesel engine supplanted steam power.

The mill buildings and tramway of St Erth Valley Alluvials during operations in the late 1920s.

Two

Marazion to Helston and North to Camborne

Another important area for copper and tin comprised the parishes of Marazion, St Hilary, Perranuthnoe, Germoe and Breage. Here again numerous small works predominated, with one or two large ones such as Tregurtha Downs Mine, Greatwork Mine, and Great Wheal Vor (wheal meaning a mine). The last was in its time the largest tin mine in Cornwall. In the area around Wendron, between Helston and Camborne, were the Polhigey and Basset and Grylls mines, while further north again was the tin mine at Parbola, or Wheal Jennings. At Roseworthy were the hammer mills and the English Arsenic Company's refinery, two important works that depended on the mining industry.

A 1920s interior photograph of part of the laboratory of the well-known firm of assayers, H.W. Hutchin & Son, which was at Tuckingmill, near Camborne.

A late-nineteenth-century view of the Tregurtha Downs tin mine, a mile north-east of Marazion. This was a very wet mine and the beautiful 80in-cylinder Cornish pumping engine, seen on the left, often had to run at exceptional speed in winter time. After the mine closed down the engine was purchased by the South Crofty Company and re-erected at their mine between Redruth and Camborne in 1903. This gave wonderful service until 1955 when it was replaced by electric pumps, being the last Cornish pumping engine to work on a Cornish mine. The engine was then preserved by the Cornish Engines Preservation Society and later handed over to the National Trust. That body had in mind a scheme for bringing the engine back to Tregurtha Downs and erecting it in its old engine house (which still stands) to form the nucleus of a working mining museum to be established there. Had this been done it would probably have been the only time in the long history of the Cornish pumping engine that one of them was erected twice in the same building! However, the scheme was abandoned and the Tregurtha Downs engine house has now been converted for residential use.

The very ancient and important Greatwork Mine worked in the saddle of high land between the summits of Godolphin and Tregonning Hills, about 4 miles west-north-west of Helston. This photograph was taken around 1890, when the mine was still working on a small scale, and shows the summit of Godolphin Hill in the distance. The ruins of the pumping engine house and stack on the Leeds Shaft, seen near the left-hand edge of the picture, can still be seen for many miles around as they are 400ft above sea level. Hamilton Jenkin has recorded that there are references to the mine as early as 1540 and as long ago as 1584 it was employing at least 300 persons continually.

Deerpark shaft at Greatwork Mine, c.1930. The sinking pump can be seen within the headgear and is about to be lowered.

Wheal Breage shaft at Greatwork Mine, *c.*1935.

The scene at Leeds shaft, Greatwork Mine, at the time of the last reworking in the 1920s and 1930s.

Lady Gwendolen Mine in the 1930s. From left to right are Lady Katherine's shaft, the tramway to the crushers, Lady Gwendolen shaft and the mill building.

Some of the tin dressing plant at the small Lady Gwendolen Mine, which was worked on the western extension of the Great Work lodes early in the twentieth century.

Wheal Vor, a mile north of Breage and 3 miles north-west of Helston was one of the greatest and richest tin mines ever discovered in Cornwall; at one time its output amounted to about a third of all the tin being mined in the county. Work probably commenced there in the fifteenth century and there were several different periods of activity, the greatest being in the years 1812-1847. In 1906 a new company was formed to drain the mine and explore the ground further eastward. The work was centred at Borlase's Shaft, using one of the first electrically driven sinking pumps to be employed in Cornwall, power being generated by the company's own steam-driven plant. The photograph shows the surface equipment as it then appeared. Unfortunately, after the mine had been largely unwatered the breakage of a single bolt led to the complete smash of the generating plant and, as the shareholders were reluctant to spend any more money on the venture, the mine was again abandoned.

An interior view of the Wheal Vor generating plant from the steam engine end of the building, c.1907.

The Wheal Vor power plant from the electrical end.

In the valley south of Wheal Vor a small plant was erected to re-treat the refuse and sand 'tails' from Wheal Vor as well as working some of the shallow parts of Wheal Metal Mine. This was officially termed Wheal Metal and Flow but was known locally as Gold Hill. This photograph showing the machinery was taken around 1910.

The 'Old Men's' Shaft at the Basset and Grylls Mine at Porkellis, near Wendron, 4 miles north-east of Helston. There have been several periods of working of this mine, the last one ending in 1938. The photograph was taken around 1908 when the mine was being reopened.

In the 1920s the Prince of Wales (later King Edward VIII) visited Polhigey Mine. The Prince of Wales is on the left and the mine manager, Mr J. Herbert Bennetts, is in white underground clothes after a visit below ground at Roberts shaft.

The headgear is erected at Roberts shaft, Polhigey Mine. The power station is on the left.

Parbola Mine, 3 miles south-west of Camborne, was also known as Wheal Jennings. An unusual type of tin deposit was worked here up until 1884 and the mine was reopened in the early twentieth century. In the latter working gas engines provided the power and the old Cornish engine house was used for other purposes. This photograph was taken around 1907.

A grinding pan was used for the fine crushing of the tin ore at Parbola when the mine was last worked. The photograph shows the large doorway of the old Cornish engine house through which the steam cylinder and other large parts of the old beam engine entered the building.

The rubber belt 'vanners' used for concentrating the tin ore at Parbola when the mine was last worked.

The Incline shaft at North Parbola Mine in the early twentieth century.

An interior view of the Roseworthy Hammer Mills, $1\frac{3}{4}$ miles west of Camborne, where the long-handled Cornish shovels were made for the mines. The wheel on the right, driven by a large waterwheel outside the building, had a number of projections around its periphery which struck the end of the rocking beam, tilting it and lifting the heavy iron hammer head at the other end of the beam. As soon as it was released the hammer fell under its own weight striking the anvil block a heavy blow. A piece of hot iron placed on the anvil could thus be beaten to the desired shape.

An exterior view of the Hammer Mills showing the waterwheels which worked the tilt hammers. Unfortunately, the whole site has since been levelled and there is no longer any indication that a most interesting small industry once existed there.

Another view of one of the tilt hammers at Roseworthy forging a piece of red-hot iron or steel, held by the smith with his tongs while his mate controls the hammer.

Standing in the fields north of the A30 at Roseworthy, a large, isolated chimney stack can be seen which is a reminder of the arsenic refinery that once existed there. The 'arsenic soot' or crude arsenic received from the mines was here burnt again with a smokeless fire and condensed in a labyrinth of brick-built chambers. When these were opened the highly poisonous, snow-white, refined arsenic could be shovelled out as shown in the picture. Note the Cornish type of miner's wheelbarrow in use.

The refined arsenious oxide was next ground to a fine powder in a machine driven by a waterwheel and passed down through a spout into small barrels, one of which can be seen on the floor beneath the man.

A general view inside the refinery showing the condensing chambers or flues on the right-hand side with iron plates luted with clay to keep in the deadly fumes while the furnace was in operation. The man is carefully sealing the barrels containing the deadly and very heavy powder. The barrel lying on its side has stencilled on its end 'English Arsenic Company'. In all these photographs it will be noted that the men use crude respirators of cotton wool inserted in the nostrils, to prevent them breathing in the fine arsenic dust, but it is said that through long exposure to it the body became inured to the poison. Nevertheless, it was essential for the arsenic men to keep very clean, washing frequently and as far as possible avoiding any sweating. Otherwise serious skin trouble could result. Safer and more controllable poisonous substances are now available and the market for arsenic has virtually disappeared, but at one time it was a valuable by-product of some of the Cornish mines.

Three

Around Redruth and St Day

Between Redruth and Camborne lies Cornwall's densest concentration of mines, with such interests as Dolcoath, Carn Brea and Tincroft, South Crofty and East Pool and Agar. This area is studied in depth in Mining in Cornwall Volume One: The Central District. To the east of the town of Redruth are the parishes of Gwennap, St Day, Chacewater and Kea, forming another very important mining district. Gwennap was famous for the Consolidated Mines (an amalgamation of numerous small mines worked with success from 1819), the neighbouring United Mines and Tresavean Mine, all large producers of copper ore. Tresavean Mine also produced tin at depth. Mining in this area goes back several centuries. East of Gwennap lies Kea, where Wheal Jane and numerous small mines are situated, while further north, near Chacewater, lie Killifreth Mine and Great Wheal Busy, the last-named an important copper producer early in the eighteenth century.

Before any steam power was available at Wheal Busy, hoisting was performed by means of a horse whim, a vertical-spindle winding drum rotated by a horse walking around a circular path. This was the way in which winding was performed in most Cornish mines when they were shallow. The picture dates from around 1907.

Car No.7 of the Camborne-Redruth Tramway at the crossing of the Portreath branch of the Hayle Railway (later the GWR) in the early twentieth century. The Railway Inn on the right is still a public house. The tramway, as well as carrying passengers, transported minerals from the East Pool and Agar mines to the mill at Tolvaddon. Mineral trains continued to use the tramway until 1934, well after passenger services had been discontinued.

East Pool Mine. The mine obtained all its coal supplies from the Carn Brea Yard of the Great Western Railway. At this point it was trans-shipped to a light railway that was constructed in the early twentieth century between the railway sidings and the mine. Prior to this, coal was hauled by horses and carts. This scene shows the small petrol-engined locomotive (constructed on the mine utilizing an engine from a Daimler motor car) with its train of side-tipping waggons and driver. The engine house contains the winding engine that hoisted from Engine shaft.

Wheal Harriett shaft at Dolcoath Mine, 1897. This new horizontal steam winding engine was built by Worsley Mesnes of Wigan, Lancashire, who tendered a more competitive price than Holman Bros of Camborne. Note the driver's operating position, which is more usual in colliery winding engines. The main shaft can accommodate two winding drums but only one is fitted here. This is because the mine shaft was only large enough for one skip road – the remainder of the available space was occupied by the pitwork of the Cornish pumping engine and ladderway.

Wheal Harriett shaft at Dolcoath Mine, c.1940. Here, the 65in engine is being scrapped. It was among the last machinery on this famous old mine to be demolished.

Robartes Engine shaft at Wheal Agar. The pumping engine is shown working for the last time, in 1924.

Headgear and plant installed on Kistle's shaft, Wheal Buller, in the 1920s.

A late 1920s photograph of Kistle's shaft, Wheal Buller, when exploratory work was carried out but a production stage was not reached.

Wheal Sparnon during the 1860s. From left to right are the beam winding engine house and stack, the hand capstan and the 70in pumping engine. The capstan shears were the highest ever to be erected in Cornwall.

Robinson's shaft at South Crofty Mine, c.1906. This photograph was taken not long after the shaft was in full commission. Pictured from left to right are the ore bins, headgear and the 80in pumping engine house, which is dwarfed by the very lofty wooden headgear. Note the tramway in the foreground, which led to Palmer's and Bickford's shafts.

New Tolgus shaft at Tolgus Mine. Pictured are the sinking headgear and bins during the time the shaft was being sunk in the 1920s. The headgear relied on guy ropes for stability, but these were found to be unsatisfactory and conventional boomstays were fitted to take the strain in the direction of hoisting.

Highburrow East shaft at Carn Brea Mine, c.1922. The 90in pumping engine which stood on this shaft had been purchased by East Pool & Agar Ltd for use at their new Taylor's shaft. Here the 'bob' or beam is being removed.

Tincroft Mine, c.1921. In the background is Harvey's 70in pumping engine, while on the right is the beam engine which formerly drove the 'man engine' and was later converted to a winding engine and hoisted from at least two shafts. Part of the horizontal compressor house may be seen on the extreme right.

Main shaft at Parc-an-Chy Mine, c.1920, taken when all the surface plant was completed during the last reworking.

The ropeway and part of the mill building at Parc-an-Chy Mine, c.1928. The mill for this mine was erected at Poldice and an aerial ropeway was constructed to convey the ore.

Seleggan Smelting Works, Redruth, c.1925. This was the last smelting works to operate in Cornwall. It closed in 1931.

The stack of Pednandrea Mine, in Redruth, still dominates the town. This photograph shows the engine shaft, c.1850. A gantry is being constructed to carry a tramway. Note the weighing machine on the right.

In 1907 the famous old Tresavean copper mine at Lanner was reopened in the expectation of finding tin beneath the once very rich and profitable deposits of copper ore, and to a limited extent this proved to be the case. Instead of installing a Cornish engine again in the old engine house it was decided to put in a steam-driven electric generating plant in the building and to drain the mine with electric pumps. In order to secure adequate draught for the large new boilers the old chimney stack was heightened to a total of 150ft – the tallest mine stack in Cornwall. The boilers stood in the low building by the side of the engine house, seen here in around 1910.

The three high-pressure Lancashire boilers for steaming the generating engines.

One of the high-speed, pressure-lubricated compound engines driving an alternator, c.1911. Later, the company took an electric power supply from the public mains and this particular engine was cleverly converted to an air compressor by changing the steam cylinders for new air compression cylinders and substituting an electric motor for the alternator.

By 1914 Tresavean Mine had been equipped with a very powerful steam winding engine, taking steam from the boilers that had previously driven the generating engines. The white winding engine house can be seen on the left of the picture. The mill and dressing plant had also been erected, as seen right of centre. The first tin was sold in 1914 and the mine continued to work until 1928, by which time it was 2,660ft deep – the second deepest mine in Cornwall. Unfortunately the value of tin fell off substantially and the mine closed in 1928.

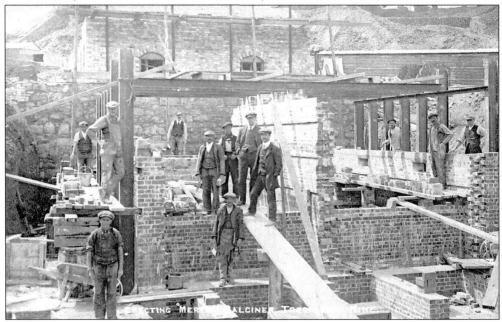

The erection of Merton Calciner at Tresavean Mine, c.1910. Calcining or burning of tin ore in a calciner produced arsenic, which was a saleable by-product. The mine would also benefit by obtaining a better price for their ore from the smelters.

Harvey's Shaft at Tresavean Mine, *c*.1928.

A water turbine driving a small set of Cornish stamps and other tin recovery machinery at the small Magdalen Mine just east of Ponsanooth, c.1913.

The electric winder house and headgear on Robinson's shaft at Mount Wellington Mine during the late 1930s. The engine house on Nangiles mine's engine shaft may be seen across the valley.

A prospecting plant for wolfram on Paull's shaft, in the South Wheal Hawke section of Great North Downs Mine, c.1944.

At the turn of the twentieth century a powerful steam-driven Cornish stamps and dressing plant was erected in the centre of the old United copper mines at Gwennap. The intention was to treat all the dumps in the area and to extract the tin which they contained. Unfortunately, with the tin recovery techniques then available this was not successful and the plant only worked for a short while. This photograph of 1902-1903 shows the rubber belt vanners of the plant, which stood within a few yards of the present car-racing track on the United Downs. The new-looking house of the beam engine that drove the stamps still stands there.

The old engine house at Shears' Shaft on the Cusvey section of the Consolidated Mines stands high on the hillside above Twelve Heads.

In 1907 the old Wheal Jane near Baldhu was reopened together with a number of adjacent old mines under the title of Falmouth Consolidated Mines Ltd. At that time the work was principally centred at Wheal Jane itself, whereas more recent operations were mostly at West Wheal Jane. This photograph, taken around 1908, shows, from left to right, the power station (partly obscured by the mill), the mill and Giles' Shaft, the main one of the mine.

The original ball mills at the Wheal Jane plant, which were not a success and were soon replaced by Californian stamps.

A large portable engine at Wheal Jane, *c.*1908. It probably supplied steam for a pump.

An interior view of the power house at Wheal Jane. The early twentieth-century reopening was Cornwall's first 'all electric' mine.

Tremayne's Shaft at Falmouth Consolidated Mines Ltd, *c.*1910. Pictured are the headgear, the ore bin and a horse-drawn mineral train conveying ore to the mill.

An auxiliary tin dressing plant owned by Falmouth Consolidated Mines Ltd in the Carnon Valley a little way below Bissoe (now the site of a rubbish dump). The photograph, taken around 1908, is interesting as it shows on the left the track of the old 4ft-gauge Redruth and Chasewater Railway and in the distance one of the famous wooden viaducts of the Great Western Railway's Falmouth branch line.

Hawke's Shaft of the Killifreth tin mine between Scorrier and Chacewater. The ruined engine house with its very tall and slender stack is still a prominent feature of the landscape. The photograph was taken soon after a short-lived attempt to rework the mine after the First World War, but the plant remained there years before being broken up for scrap during the Second World War. This mine is generally regarded as being one of the best prospects for tin in Cornwall.

When it was proposed to reopen the Great Wheal Busy copper, tin and arsenic mine near Chacewater in 1907 there was great rejoicing in the village, for it meant much work for a district where there was not much employment. A bullock was given by Lord Falmouth and roasted whole as part of the festivities.

A general view of the Wheal Busy milling plant containing twenty-five head of Californian stamps and tables, etc., and nearer to the camera, the arsenic roasting plant, condensing chambers and chimney stack, c.1908.

The primary purpose of reopening Wheal Busy was to obtain a ready supply of arsenic for an Anglo-Belgian chemical manufacturing company. It was thus necessary to install, among other machinery, a calciner for roasting the arsenical ore and brick condensing chambers and flues leading to a tall chimney stack. This photograph shows the stack at its completion, around 1907.

Originally it had been intended to unwater Wheal Busy by means of an electric pump. A gas engine-driven generating plant was therefore installed to provide the power. The early type of centrifugal pump employed was, however, a complete failure and it was therefore decided to erect a powerful Cornish engine in the old engine house, which still stood there from a previous working of the mine. The engine, which had originally worked on a Cornish mine near Par and had then gone to a colliery in South Wales, was purchased and brought back to Cornwall in 1909 for re-erection at Wheal Busy. Here, the massive 35-ton beam, made of rolled iron plates about 4in thick, is inched towards the house prepared for its reception.

The engine house at Wheal Busy with the beam up in place showing the big pulley blocks used for lifting, and the timbers up above that bore the weight while the lift was in progress.

The Wheal Busy 85in-cylinder pumping engine completed and at work in 1910. On the left is the new boiler house containing three Lancashire boilers and on the right a part of the balance bob. Note the man in the kibble high above the beam. It is a point of interest that it was at Wheal Busy in 1778 that James Watt erected his first pumping engine in Cornwall – the precursor of the true 'Cornish beam engine' as it came to be known all over the world.

Unfortunately, the Anglo-Belgian Company appears to have always been short of capital and the working of the mine was an intermittent affair. Finally, the plant was taken over by the adjoining Killifreth Company and in 1923-1924 Wheal Busy was again active for a short while as arsenic was then fetching a high price. This was the last time that the fine old pumping engine worked and shortly after the end of the Second World War she was broken up for scrap after having stood in the house unused for so many years. Here, in 1952, the lower part of the cylinder has been smashed with dynamite, the piston rod has been blown to one side and the piston rings lie on the floor.

Four

Around St Agnes and Perranzabuloe

North of Redruth, in the parishes of St Agnes and Perranzabuloe, a broad strip of country contains a large number of small mines, many of considerable antiquity. Among the more important concerns were Wheal Kitty in St Agnes, West Wheal Kitty and Blue Hills Mine. These were all worked for tin in the nineteenth century, while Wheal Kitty worked as late as 1930. Further east was West Chiverton Mine, which worked for lead, silver and zinc. Numerous small workings, some of which worked into the twentieth century, are to be found around Perranporth, while the Great Perran Iron Lode was worked at Treamble, where the huge excavation is now used as a caravan park.

One of the big tractors and scrapers employed on the Great Perran Iron Lode opencast project in 1938.

Plant under erection, *c.*1901. At the turn of the twentieth century a great advance in ore dressing was achieved by the development of the flotation process by which a good proportion of the world's minerals are now recovered. It is therefore a matter of considerable historic interest that one of the earliest experimental flotation plants, using the Elmore oil vacuum process, was erected at the Tywarnhayle copper mine near Porthtowan.

Part of the Tywarnhayle Mine viewed from across the valley, *c.*1860. At the top of the picture is the Gardiners shaft beam winding engine house and stack, as well as the 80in pumping engine. To the right of the cottages is the copper crusher engine and buildings. In the foreground is the rim of the famous 'Navvy Pit', formerly an underground mine named Wheal Music.

The old waterwheel-driven corn mill at Porthtowan. Many miners were also small farmers or fishermen and it was little mills such as this which served the needs of the community in the days when mining was at its height throughout the length and breadth of Cornwall.

Wheal Noweth. At the end of the Second World War a former Wheal Kitty miner named Joe Yelland and his son Martin (pictured) commenced sinking a shaft at Mingoose, St Agnes. This was situated on the downs above East Wheal Charlotte. Their first windlass was a domestic mangle, as can be seen in the picture!

East Wheal Charlotte, or Dale's Mine, was a small tin mine on the eastern side of the Chapel Coombe valley, which runs inland from the coast at Chapel Porth, near Porthowan. This must not be confused with the much larger copper mine, sometimes referred to as Great Wheal Charlotte, which is on the cliffs just west of Chapel Porth. The little mine shown in this photograph was being worked around 1912, partly by means of an adit level driven into the hillside and also from a flat inclined shaft sunk further up the hill. Note the small steam-driven set of Cornish stamps and the round 'buddles' for separating the tin from the waste.

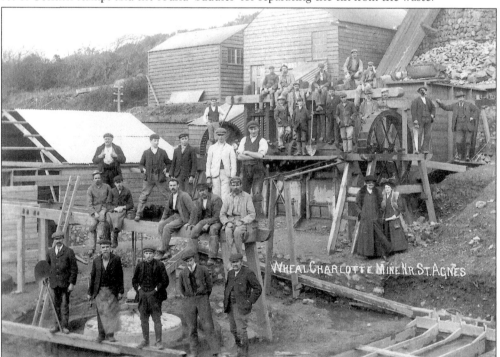

East Wheal Charlotte, or Dale's Mine, with the surface employees, a couple of miners and two lady visitors gathered around the plant, c.1910. A typical example of the so-called 'good old days' when everything was simple, when most men could turn their hands to a number of trades and there was the minimum of paperwork. Wages were low, but so was the cost of living in Cornwall!

The Cornish stamps beam engine at the Wheal Kitty mine at St Agnes whose house was destroyed by fire in 1905; it was thought to have been due to arson by an official of the mine who had been sacked. The house was rebuilt and the engine worked again for some years. This photograph shows clearly how the power from the steam engine was transferred from the steam cylinder on the left to the crank and flywheels on the right via the overhead rocking beam or bob, which pivoted on bearings on top of the massive bob-wall – virtually the only part of the engine house left standing.

An old and damaged, but interesting, photograph showing the building of the pumping engine house at Sara's Shaft of Wheal Kitty in 1910. Note that although the roof timbers are in place the brick upper part of the stack has not yet been built. Most unusually, the beam of the engine is already partly in the house with its rear end protruding through the wide 'cylinder opening' or large doorway. This was the last Cornish engine to work in the St Agnes district and only stopped in 1930.

Miners leaving the cage in Sara's Shaft at Wheal Kitty, c.1919. Note on the right two large 'kibbles' for raising ore and stone.

Wheal Kitty, c.1928. In consequence of difficulties arising from the First World War Wheal Kitty had to suspend operations in 1919, but the mine was reopened in 1926 and much new plant installed. On the left is the new mill, in the centre Sara's Shaft and pumping engine and on the right the winding engine and new compressor and boiler house, new stack etc. A new and rich tin lode was cut at the bottom of the mine but it was closed down in the worldwide slump of 1930.

Sara's shaft at Wheal Kitty, St Agnes during the 1930s. The engine house had contained a 65in Cornish pumping engine. Here one of the boilers is removed.

West Wheal Kitty, on the western side of the Trevaunance Valley, was one of the richest and most profitable tin mines ever worked in the St Agnes area. Here, a group of West Kitty miners are pictured outside their 'dry' or change house. Note the large bunch of miner's candles carried by the man on the left. This was probably for a large 'pare' (or gang) of men working together in one part of the mine.

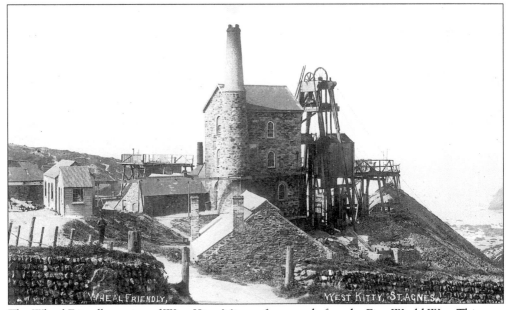

The Wheal Friendly section of West Kitty Mine, a few years before the First World War. This now roofless engine house still stands high above the deep Trevaunance Valley on its western side.

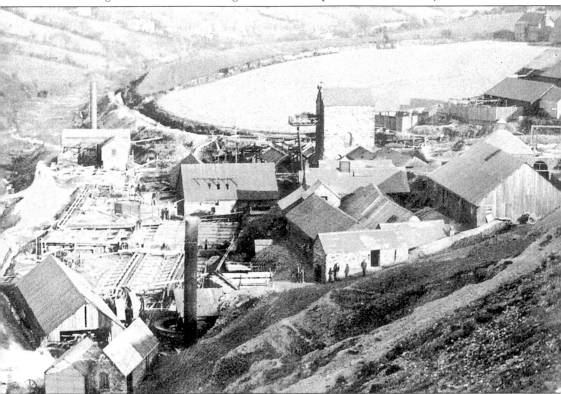

Both the old and new steam-driven stamps and tin-dressing plant of West Kitty at 'Jericho', about a mile east of St Agnes. All the ore from the mine which was worked under the town had to be transported here by horses and carts.

The seaward end of the Trevaunance Valley at St Agnes. The old buildings are said to have been originally an iron foundry, possibly the Trevaunance Foundry which is recorded as having been sold in 1865. Later the place housed a part of the tin recovery plant seen in the photograph. In the background is the old St Agnes harbour where most of the coal used in the local mines was brought in by means of small sailing vessels.

Long after regular mining ceased at the old Polberro Mine at St Agnes tributers continued to work there on a small scale above water level. Shown is a 'horse whim', used for raising ore to the surface, around 1910. (Photograph: Albert J. Fellows)

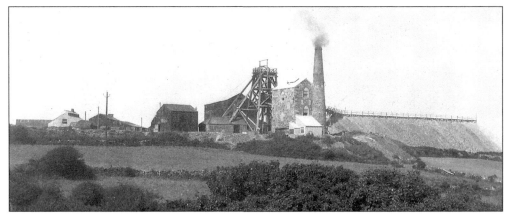

The most recent mine to work in St Agnes was the Turnavore Shaft of Polberro Tin Ltd, which was reopened in 1937 and deepened from 650ft to 1,090ft, making it the deepest shaft in the St Agnes district. This work was done in the hope of intersecting the rich West Kitty and Wheal Kitty lodes which were abandoned when Wheal Kitty was closed down in the world-wide slump of 1930. The first-named lode was indeed cut and a good deal of work done on it, but the acute man power and other difficulties which developed during the Second World War prevented output from being built up to an economic level. Although tin was then being produced it became impossible to continue and the mine was closed early in 1941. This was a great pity for soon afterwards the Allies lost the great eastern tin-producing countries and then every ton of tin that could be mined in Cornwall was of the utmost importance to Britain's war effort. In this, the last working of the Turnavore Shaft, the plant was all electrically driven, but the old engine house was converted into a miner's change house. The smoke from the old stack came from a small heating boiler used to provide hot washing water etc., and not from large steam boilers as previously.

A fine photograph of the old Blue Hills Mine at the seaward end of the Trevellas Valley, St Agnes, c.1890. In the centre is the 70in-cylinder Cornish pumping engine at the Engine Shaft and, to the left, the big flywheel of the horizontal single-cylinder engine driving the stamps and also winding from several shafts.

A strange sight even in a land of strange things! The remains of the pumping and winding machinery of a small lead prospect in the shallow valley a mile north of Mithian Church. It was known as Wheal Treasure or the Silverwell Lead Mine, but it was also cynically remarked that 'It did not contain enough lead to make earrings for a black beetle!' Apparently the old iron chimney (the fire tube from an old boiler) did not provide sufficient draught for this boiler and it was therefore heightened with the top-heavy brickwork seen in the picture. Even though two out of three supporting chains had rusted away, the old stack withstood all the gales and was only demolished during the scrap iron drive of the Second World War. The mine was last worked before the First World War.

A large and most unusual engine house at Batters' Shaft of the famous West Chiverton lead and zinc mine, half a mile north of the A30 and $1\frac{1}{4}$ miles west of Zelah Village. Instead of the stack being built at one of the back corners of the house it was built on the centre line, thus completely blocking the back of the building for bringing in the 80in-cylinder and beam. The result was that the cylinder opening, or large doorway, had to be in the side of the house (making this, with so many other openings, a weak wall) and the heavy beam probably had to be lifted in at the front end with the weight suspended from the headgear. The whole thing must have been a very awkward erection job and quite unnecessary as there was plenty of room behind the building to bring in the heavy parts if the house had been built in the normal manner. West Chiverton was at one time the most productive and profitable of the Cornish and Devonshire lead mines and employed 1,000 people. It closed down in 1886.

Batters' engine house from the western or boiler house side. The two small wall openings at the lowest part of the building were for the cylinder steam jacket pipe on the left, and the main steam pipe on the right. The tall doorway still further right was to give access from the boiler house to the driving floor or 'bottom chamber' of the engine house.

A very early photograph of Perranporth, showing the derelict engine houses of Wheal Leisure and the other extensive copper mines in the area. Even the shape of the small island known as Chapel Rock has changed since this photograph of c.1850.

Lambriggan Mine. This plant was installed during the 1920s when the main shaft was deepened.

A general view of the main shaft at Lambriggan, *c*.1927, during the reworking.

Cligga Mine, Perranporth, *c*.1930. The headgear on the left served the inclined tramway on the cliff which was operated by a steam winch in the building to the right. Contact shaft winding engine house and headgear complete the scene.

The 300ft-high cliffs on Cligga Head, showing the portals of the levels of Cligga Mine, c.1930. The inclined tramway can be seen on the left.

The tin and wolfram mine at Cligga Head, a mile west of Perranporth, was worked on quite an extensive scale during the Second World War. Here the headgear and mill buildings are erected in 1941-1942.

Prospecting operations at Bolingey, c.1910. This led to the opening-up of the Wheal Alfred zinc mine.

Wheal Vlow was situated at Perranporth and reopened in the 1920s. The print shows the Count House, smith's shop and other buildings. In the background, from left to right, are the power house, winding engine house and headgear on Hallett's shaft.

About two miles north-east of Perranporth there is a large vein of ore known as the Great Perran Iron Lode which was worked extensively in the 1870s. Shortly before the Second World War an attempt was made to work it again at Treamble by opencast methods using the recently introduced bulldozers and scrapers. Two of these machines can be seen at work in the picture. The venture was not an economic success for the great vein was too irregular in shape to permit working in this way, but some iron ore was obtained during the wartime emergency by using more conventional underground methods.

Five

Around St Austell

One of the two productive uranium mines in Cornwall was at South Terras, at the western end of the St Austell mining and china clay area. The district was one that had only a few large mines and some associated industry, including the smelting works with the tallest stack in Cornwall. With the expansion of the china clay industry, mining in the area declined in importance. Stream and shallow surface workings survived to the north, on Goss Moor, into the twentieth century and a small wolfram mine worked successfully, but not continuously, between 1918 and 1958 at Castle an Dinas.

One of the great horse-drawn wagons previously used in the china clay industry.

South Terras Uranium Mine, *c.*1920. This picture shows the headgear and the tramway to the dump.

Ventonwyn was a small tin mine half a mile north west of Hewas Water on the Truro to St Austell road, and the old engine house is a prominent landmark visible from several miles away. This rotative beam engine did the pumping in addition to driving the small battery of Cornish stamps. The interesting thing about this photograph (around 1910), however, is that every man, whether surface or underground, is shown exhibiting the tools of his trade. For example, the fourth man from the left in the front row is the pitman, holding a Cornish pump 'bucket'. The man next to him is the sample tryer, holding his 'vanning shovel'. The seventh and eighth men from the left (standing) are the blacksmiths with their hammers. The man on the extreme right with the oilcan in his hand is undoubtedly the engine driver. The men with candles fixed with lumps of clay to their hats are, of course, the miners.

Steel headgear and crusher station being erected at Dowgas Mine in the early twentieth century.

In Cornwall's big china clay industry, in the days of steam power, it was usual to have a Cornish beam engine pumping from each pit. A separate horizontal engine would haul the waste material up the inclines to the great conical tips that were such a prominent feature of the landscape. Since the beam engines varied their length of stroke with the slight variations of steam pressure in the boilers that occurred when the winding engines were working, it was very difficult for one driver to look after two engines simultaneously. A local engineer, 'Jackie' Menire, therefore invented what was known as the 'St Austell governor', for automatically controlling the length of piston stroke of the beam pumping engines. Pictured is one of these simple but clever devices as applied to the 50in-cylinder pumping engine at the Parkandillick clay works at St Dennis. Fortunately, this engine was preserved in working order by her owners, English Clays Lovering Pochin & Co Ltd., now E.C.C. International. (Photograph: W.J. Watton)

Carpalla clay works' 40in-cylinder engine, south of Foxhole. This engine had worked previously on Thomas's Shaft at the celebrated West Wheal Kitty at St Agnes. In 1915 it was removed to Carpalla, being the last beam engine to be erected in the whole of the central clay area of Cornwall. It worked there until replaced by electric pumps in 1944 and was later dismantled and placed in store in London with the intention that it would eventually be re-erected at the Science Museum. This little engine, of classic form, was built by Harvey & Co. of Hayle, probably the most famous of the Cornish engine building firms.

Bloomdale clay works pumping engine near Foxhole in 1935. This was a much-rebuilt 36in-cylinder engine, originally made by the Perran Foundry Company. Previous to being at Bloomdale it had worked at the Polbreen Mine, St Agnes. The man who drove this engine was noted for his encyclopaedic memory for dates. There appeared to have been no local or national happening that he could not give the precise date for no matter how many centuries ago the happening. His memory even extended to most foreign countries; he really was extraordinary! (Photograph: Frank D. Woodhall)

The 50in-cylinder engine at Goonvean clay works, a mile north of St Stephen, *c*.1929. This engine was made by Harvey & Co. in 1863; after working at two mines in St Agnes it was moved to its present site in 1910. In 1928 the beam broke and a new and heavier one was cast by Holman Brothers Ltd of Camborne. This was probably the last bob made for a beam engine anywhere in the world. This engine ceased working several years ago but it is still retained in its house by the owners, the Goonvean and Restowrack China Clay Co. Ltd.

The last beam pumping engine to work commercially in Cornwall was the 30in-cylinder engine at the Greensplat clay works, about 2 miles north-north-west of St Austell. This delightful little engine was originally a rotative one but its flywheel was purchased for another engine at South Greensplat works. It was then decided that, instead of scrapping the remainder of the engine, it should be used as a non-rotative pumping engine and as such it was re-erected at Greensplat in around 1887, and it continued to work there until 23 February 1959. It bears neither the maker's name nor date, but it may have been made by the St Blazey Foundry. The engine remained in its house at Greensplat until 1972 when it was purchased by the Wendron Forge Museum (near Helston) and re-erected as an open-air exhibit where it can be seen working – the actual movement now performed by means of an hydraulic ram.

The 22in-cylinder rotative beam engine at the Great Hallaze clay works about 2 miles north of St Austell, just before the Second World War. This little engine originally worked at the Devon Gawton copper and arsenic mine on the Devonshire bank of the Tamar River, but in around 1910 it was bought and re-erected at the new clay works. This was a failure, however, and the engine did not work long there and lay idle in its house for many years. Shortly before the Second World War it was decided to connect this clay pit with another one by driving a tunnel between the two, thus forming a capacious reservoir for public water supply. Electric pumps were at first employed to lower the water level in the pit to enable the tunnel to be driven, but the water was so gritty that there was continual trouble with the pumps. It was then decided to try to work the old engine again and, as rusty and derelict as she appeared to be this was a great success and enabled the tunnel to be driven. Sadly, after the job had been completed the old engine was broken up for scrap – she deserved a better fate!

At Great Hallaze most of the woodwork and floors within the house had gone and temporary planks had to be laid down to enable the driver to move around the engine. A ladder was used to oil the 'nose' of the bob and all the boiler fittings were replaced; the boiler house had been used as a cattle shed and was half full of hay and straw. No modern diesel or electrical plant would have worked again if they had been subjected to such neglect as had the old steam engine.

The famous Par Stack of the silver-lead smelting works owned by the Treffry Estate. It was built around 1860 and is variously stated to have been 235ft or 280ft high. Whichever dimension is correct, it was undoubtedly the tallest stack built in Cornwall and was said to contain a million bricks. At the time of building it was fashionable for gentlemen to wear tall top hats and these became known in Cornwall as Par Stacks. The base of the stack stood right by the side of one of the Great Western Railway's tracks and the company was suspicious of the strata on which it was built. Therefore, as it had long been disused, they decided to demolish it in 1907 and in the photograph the steeplejacks hold aloft the first bricks removed from the top. After it had been reduced to a part of the original height the remainder was thrown towards the sea by an explosive charge on 23 August 1907. In consequence of its demolition a modern landmark was set up to guide shipping as the great chimney had done for so many years.

The Goss and Tregoss Moor, north of the great central china clay area, is the largest alluvial tin deposit in Cornwall and has been worked for tin from time immemorial. Unfortunately, the tin which remains is difficult to recover and a company which was formed to rework these alluvials burnt its fingers badly when it commenced dredging there in April 1925. The new dredge, pictured, was 80ft long with a beam of 35ft and cost £15,000 (then a sizeable sum), but it only worked for a short while and was eventually cut up for scrap.

North shaft at Castle-an-Dinas Mine, c.1938. The Castle-an-Dinas wolfram mine was worked beneath the hill of that name, north of Goss Moor, and was a most productive and profitable mine although it only contained one narrow lode or vein. The original workings were on the north side of the hill. Pictured are the steam hoist house, headgear and dump.

By 1941 the mine had extended so far south as to necessitate the sinking of a new shaft on the southern slope of the hill. Here, the final raise has been brought through to surface and the shaft collar is being formed. The figure third from left is the late Mr Joe Chynoweth, mine manager during the 1940s.

South shaft at Castle-an-Dinas Mine, c.1942. Pictured are the steam winding engine (previously at King Edward Mine) and the boiler, with the house yet to be built around them. The headgear is being erected.

South shaft at Castle-an-Dinas Mine, c.1942. Pictured from left to right are the steel stack, the winder house, the base of the crusher station, the headgear and the combined pumping engine and compressor house.

This unusual photograph was taken from the top of the headgear of the new shaft at Castle-an-Dinas looking down on the big Lancashire boiler and the winding engine during erection in 1942.

South Shaft. An early 1940s view from the top of the headgear showing the decking from the landing brace (track yet to be laid) to the crusher station. The aerial ropeway terminal is under construction.

South shaft at Castle-an-Dinas. This unusual 1940s view looks upwards from just below the shaft collar and shows the noses of the Cornish pitwork's twin 'bobs'.

Again an early 1940s view – this time a photograph taken from the crusher station of the pump 'bobs' and rods during installation. The figure is the late Mr Walter Langford, well known in mining circles over many decades. He was in charge of the installation of the shaft pumping arrangements.

One of the open cavities or stopes at Castle-an-Dinas Mine after the ore had been extracted. The timber was not required for support of the walls of the stope (which were strong and firm), but for the stage planks on which the miners worked while they drilled their holes for explosives. This was a method of working peculiar to this particular mine and one which enabled the broken ore to be sent to the mill day by day, as it was broken. What appears to be a ball of light by the miner's arm is actually a powerful miner's lamp facing the camera.

Six

East to the Tamar

Among the open works to the east of St Austell was Great Treveddoe Mine, now almost completely obscured by vegetation. Around Caradon Hill, in the parishes of Linkinhorne and St Cleer, were numerous mines, including the Phoenix Mines., Redmoor Mine, north of Callington was worked by a number of different companies in the nineteenth century, while east of it lay Kit Hill Consols, on the hill of that name, and Hingston Down Consols, which later worked with Clitters Mine. The Clitters United company was one that was successful at working complex ores containing several commercially valuable minerals, though the earliest plant built to do this type of ore-dressing was erected over a century ago at New Consols Mine, some two miles to the north west. On the south slope of Hingston Down lay the Prince of Wales mine, whose shares, early in the twentieth century, were a popular gamble among the French admirers of King Edward VII. Nearer to Devon lay Okel Tor Mine, which was worked from the mid-nineteenth century, at first for silver-lead, then for copper and finally for tin and arsenic.

A small electric winding engine at South Phoenix Mine, c.1908.

Great Treveddoe Mine near Warleggan was an old tin and copper mine worked below the floor of a massive open quarry known as Old Wheal Whisper. On average, the grade of tin ore here was always low, but as the plant was driven by water-power it was possible to work the mine at very low cost.

An old waterwheel-driven set of Cornish stamps, believed to be at Great Treveddoe Mine.

The Phoenix United Mines, 1 mile north-west of Caradon Hill, closed down in 1898. After producing a great deal of copper these mines became the largest tin producers in the eastern half of Cornwall and in their day were very profitable. In 1907 the Phoenix Mines were most unwisely reopened for it should have been obvious from the records of the nineteenth-century company that the mines had become poor in depth. However, it was decided to sink a fine new perpendicular shaft equipped with a new 80in-cylinder pumping engine and the photograph shows the building of the new engine house and the erection of the temporary headgear for the sinking of the shaft.

The Prince and Princess of Wales (later King George V and Queen Mary) visited Phoenix in June 1909 when the Prince christened the new shaft 'The Prince of Wales Shaft' and the Princess started the new Cornish pumping engine – the last one of its size to be built anywhere in the world. The shaft was sunk to a depth of 1,200ft; the mine drained to the bottom and a certain amount of development done below the bottom of the old workings. This, however, only exposed small patches of good ore and the mines were again abandoned in 1914. Pictured is the new plant at the Prince of Wales Shaft as it appeared in around 1909. On the hillside in the background can be seen the Cheeswring granite quarry with the famous Cheeswring pile of rocks on the skyline on the edge of the quarry.

Phoenix United Mines, c.1908. In the last reworking, which ceased in 1914, a Holman-built 80in pumping engine was erected on the new Prince of Wales shaft. Here the three Lancashire boilers that were to provide steam for the engine are installed.

The cover of the 80in-diameter-cylinder of the new Phoenix pumping engine built by Holman Brothers Ltd of Camborne in 1907. After the mine closed again in 1914 the engine was greased over and remained in its house until around 1931 when it was, most regrettably, broken up for scrap. This is said to have been a good engine and would run at speeds up to nine strokes per minute with ease.

The unusual steel plate beam of the new Phoenix Mines' pumping engine in 1909. Unfortunately, the bright sunlight coming through the window over the beam has slightly spoilt the photograph.

The fine new steam winding engine built by Holman Brothers Ltd in 1907 for the Prince of Wales Shaft. When they closed down again this engine was sold for further service at a colliery.

A general view of the new mill and Prince of Wales shaft with a high wooden headgear at Phoenix United Mines, c.1912.

When the Phoenix Mines were last worked the old custom of the mine barber was revived whereby for a small monthly sum of money the men could have a haircut and a weekly shave.

Immediately east of Kit Hill and within a mile and a half of the River Tamar is the high ridge of land known as Hingston Down. At the highest point, 880ft above sea level, is the Hingston Down Consols tin and copper mine which was reopened in the early twentieth century in conjunction with the Gunnislake Clitters Mine on the Cornish bank of the Tamar. To pump the water at Hingston Down Mine a rotative beam engine was purchased from the Devon Great Consols copper and arsenic mine which had then recently closed. Here is the new engine house in 1903 at Hingston with the bob or beam up in place, and the big flywheel being reassembled. With the aid of a strong glass the name of the maker of the engine can be read on the bob on the original photograph, namely 'Bedford Foundry, Tavistock'. This engine house, with its roof hidden by parapet walls, still stands as a prominent landmark on the summit of Hingston Down.

Hingston Down Consols seen when working in the early years of the twentieth century. Unfortunately, the shaft in which the pump worked proved to be crooked and a great deal of trouble was experienced with the plant. Electric power was then generated in the buildings seen on the right of the picture but the early type electric pumps employed also gave a lot of trouble and in the end it was the Cornish pump on which the company had to rely.

The mill for the treatment of the ores from the Hingston and Clitters Mines was on the steep hillside immediately below the latter mine and above the Tamar, which can be seen in its twisting course on the right.

The plant in the Clitters mill, c.1903.

Tables, etc., in Clitters mill.

At the village of Luckett, $2\frac{1}{2}$ miles further up the Tamar valley than the site of the Clitters mill, there is an old and interesting mine that produced a good deal of copper and arsenic ores and a certain amount of tin. It has been variously named Wheal Martha, New Consols and New Great Consols, and had a varied but intermittent career between 1838 and 1952. Although the dumps were worked on a considerable scale during the First World War for tin, wolfram and arsenic, underground operations ceased as long ago as 1879, leaving a mass of old-fashioned mining machinery rusting away until it was broken up for scrap just before the Second World War. Here are some of the old engines as they appeared around 1938 shortly before being scrapped. Pictured from left to right are the 80in-cylinder pumping engine on Phillips' or the Engine Shaft, the remains of a rotative beam engine which drove the Cornish rolls for crushing the copper ore and the beam winding engine whose house was nearly covered with ivy. This was so old that it had chains on its drums instead of wire ropes. It has been said that the whole mine was an industrial site of unique interest and it is a great pity that this machinery was scrapped. In 1946 the mine was reopened and drained to the bottom by means of electric pumps. A further 170 tons of black tin (i.e. tin oxide) and a little wolfram were produced, but it was not a successful venture and the mine closed again in December 1952.

The very old Cornish pumping engine and its most unusual house at New Consols, 1938. The building originally contained a 50in-cylinder engine; when more power and an altogether larger engine (which had a different length of stroke) was needed the house was not suitable. Furthermore, because of the position of the chimney stack at the back of the building, the large doorway through which the cylinder enters could not be sufficiently enlarged to bring in the new 80in cylinder. In order to avoid the cost and time of building an entirely new engine house the difficulty was ingeniously overcome – the thick 'bob end' (i.e. the thick wall which bears the weight of the rocking beam) was demolished and replaced by a thicker wall at the requisite distance from the mine shaft. As the new masonry might not bond sufficiently strongly with the old, the new wall was extended with a buttress on each side, so as to give it extra strength, and these side extensions were covered with small gable roofs. Furthermore, the wall opening for the 'plug-door' under the beam, which is always made as small as possible so as not to weaken the wall under the weight of the beam and its load, was made sufficiently large to enable the large cylinder to be brought into the house that way. After the cylinder was in place a false arch of brickwork was built inside the large opening thus strengthening the whole wall. Here the dismantling of the engine has just commenced.

At New Consols, in addition to all the other engines, there was on the southern side of the valley a large Cornish stamps engine – complete with the stamps – and several calciner furnaces for roasting the arsenical ores. Pictured is the stamps engine in the 1920s with its two odd size flywheels and the auxiliary beam for pumping the 'dressing' or processing water required for the treatment of the tin ore. The engine itself was of unusual design with cam and roller valve motion and with the valves at the side of the cylinder instead of in front as is usual. In common with all the other engines on the mine, this one still had its old Cornish boilers in situ, but in each case the boiler-house roof had collapsed.

Headgear and portable air compressors pictured during the early period of the late-1940s reworking of Engine shaft at New Consols Mine, Luckett.

Headgear erected during the mid-twentieth-century reworking of Engine shaft at New Consols Mine. The derelict house of the Cornish pumping engine from an earlier working is on the left.

The new mill and aerial ropeway constructed during the reopening of New Consols Mine in the late 1940s and early 1950s.

Watson's Shaft at the Prince of Wales Mine at Harrowbarrow, on the southern side of the Hingston Down ridge. This early twentieth-century photograph shows the interesting 'hollow-work' or trellis-like cast iron bob of the 50in-cylinder pumping engine. This type of engine beam was rather rare and this was the last of the type to work in Cornwall – the mine closed in 1916. Note also the old type of portable steam engine on the left pulling a skip up an incline with material to fill an old working that had caved in from down below and was in dangerous proximity to the engine house. The automatic tipping arrangement of the skip is also interesting.

Watson's shaft at the Prince of Wales Mine, Harrowbarrow, in the late nineteenth century. Pictured are the pumping engine house, headgear and winding/capstan engine. Note the vast sand dumps.

A close-up view of the pumping engine and headgear on Watson's Shaft at the Prince of Wales Mine. The hollow-work bob of the engine can be clearly seen, as can the timber surface balance beam. The bell high above the engine house was a survivor of earlier days before the steam whistle or hooter was in use to tell the men when the shifts began or ended – few owned a watch in those days! A mine in Cornwall was called a 'bal' and hence this bell on every mine was referred to as the 'bal bell'. The Prince of Wales Mine was owned latterly by a very impecunious company and its abandonment was due to special circumstances. It is generally thought to be a good prospect for tin if worked on modern lines and with a different prevailing metals market from that today (1999).